ISBN 978-1-330-29730-8
PIBN 10017789

1 MONTH OF
FREE
READING

at

www.ForgottenBooks.com

By purchasing this book you are eligible for one month membership to ForgottenBooks.com, giving you unlimited access to our entire collection of over 1,000,000 titles via our web site and mobile apps.

To claim your free month visit:

www.forgottenbooks.com/free17789

English
Français
Deutsche
Italiano
Español
Português

www.forgottenbooks.com

Mythology Photography **Fiction**
Fishing Christianity **Art** Cooking
Essays Buddhism Freemasonry
Medicine **Biology** Music **Ancient
Egypt** Evolution Carpentry Physics
Dance Geology **Mathematics** Fitness
Shakespeare **Folklore** Yoga Marketing
Confidence Immortality Biographies
Poetry **Psychology** Witchcraft
Electronics Chemistry History **Law**
Accounting **Philosophy** Anthropology
Alchemy Drama Quantum Mechanics
Atheism Sexual Health **Ancient History**
Entrepreneurship Languages Sport
Paleontology Needlework Islam
Metaphysics Investment Archaeology
Parenting Statistics Criminology
Motivational

THE YERKES OBSERVATORY

By Edwin Brant Frost, Director

The establishment of a large observatory for astrophysical research in connection with the University of Chicago was the conception of George E. Hale, who had achieved success in his solar work with his private equipment in Chicago known as the Kenwood Observatory. President Harper took an active interest in the realization of the plan, which presently took shape as an opportunity occurred for the purchase of two very perfect glass disks (crown and flint) of 42 inches diameter, sufficient for the construction of an object-glass of 40 inches clear aperture. These disks had been cast by the well-known makers of optical glass, Mantois of Paris, upon order of an institution in southern California, which was, however, then unable to raise the money for figuring and mounting the telescope. The opportunity was presented by Dr. Harper and Mr. Hale to Mr. Charles T. Yerkes, a keen business man of Chicago, of purchasing these disks and having the largest refracting telescope in the world. Mr. Yerkes agreed in September, 1892, to finance the undertaking, and a contract was made with the famous firm of Alvan Clark & Sons, of Cambridgeport, Mass., for figuring the two disks, which were then in their hands, and with Warner & Swasey, of Cleveland, Ohio, for constructing a suitable mounting.

Mr. Hale was appointed director of the new institution, and gave much study to the plans for the instruments and building. He outlined in October, 1892, the principal lines of scientific work which would be undertaken.

The site of the observatory was selected after a careful investigation of the available locations and after an expression of opinion had been obtained from numerous astronomers familiar with large telescopes. On account of the smokiness, dust, and jar, as well as the glare due to the lights of a great city, it was at once obvious that the observatory ought not to be established within or near the limits of Chicago. But by reason of its connection with the University, it was practically essential that it should not be placed more than one hundred miles away from the other departments of the University. Tracts of land were offered in more than twenty places in Illinois and elsewhere, but the situation at Lake Geneva seemed to offer the advantages of freedom from disturbance by commerce, present or prospective, and from the encroachment of too numerous dwellings; together with sufficient accessibility from the University and city. The site of 53 acres was presented by Mr. John Johnston, Jr.; other adjacent land was later acquired by the trustees of the University, so that since 1907 the grounds have included about 70 acres, with a frontage of 600 feet on the lake (where a pier for steamers is maintained in summer).

The Observatory is situated about 190 feet above the level of the lake, or 1,050 feet above the sea. Its geographical position, as determined by officers of

3

FIG. 2.—THE 40-INCH REFRACTOR, TAKEN WITH THE RISING FLOOR NEARLY AT ITS
LOWEST POINT

the United States Coast and Geodetic Survey in 1900, is: Latitude, 42° 34′ 12″64; Longitude, 5ʰ 54ᵐ 13ˢ24 west of Greenwich. The building is one mile from the post-office of the little village of Williams Bay, Wis., and is a mile and a quarter from the station, at the terminus of a branch of the Chicago & Northwestern Railway, 76 miles from Chicago.[1]

In the summers of 1907 and 1908 a beginning was made in grading the grounds of the Observatory after a plan designed by Olmsted Bros., of Brookline, Mass. When this plan is completed and the planting has been carried out as proposed, an appropriate setting will be secured for the architecturally impressive building.

The Observatory was designed by Henry Ives Cobb, who carried out the plans drawn up by Professor Hale in a way that successfully combined architectural effect with scientific utility. The large observatories of Europe and America were visited by Mr. Hale before the plans were begun; the most useful suggestions being received from the designs of the Lick Observatory and the Astrophysical Observatory at Potsdam, Prussia. The style of the building is Romanesque with rather elaborate details. It is constructed of brown Roman brick, with terra-cotta ornaments to match. The shape is that of a Latin cross, with the three towers and the meridian room at the extremities. The long axis lies east and west, with the great tower at the west, 92 feet in diameter. The entire length of the building in this direction is 326 feet. The two smaller domes are on the north and south axis, with their centers 144 feet apart.

The basement floor contains the instrument shop, the optical shop, a carpenter shop, a large physical laboratory adjoining a room fitted with a Rowland concave grating, and several dark rooms for photographic work. The main floor contains seven offices for the staff, a laboratory, computing room, and lecture-room, a reception room for women, a library 40×20 feet, besides dark rooms, chemical laboratory, and cabinets for instruments. The low attic along the east and west axis is available only for storage, but an important room is provided on the second floor between the smaller towers, called the heliostat room, nearly 100 feet long. At the north end the roof can be slid back so that a good view of the sky is obtained, and a heliostat can be mounted on the pier at that point. The horizontal beam of light can then be studied in the particular manner desired in the long room, which is provided with numerous piers. The most important piece of work thus far done in this laboratory was the first successful measurement of the heat of the stars by Professor E. F. Nichols (then of Dartmouth College) in the summers of 1898 and 1900.

THE GREAT REFRACTOR

The forty-inch telescope is mounted upon a massive brick pier, which rests upon a solid concrete foundation set in the prevalent gravel formation. The column is of cast iron in four heavy sections. The center of motion of the telescope is 62

---

[1] Additional railway facilities are obtained by the trolley line terminating at the head of the lake (Fontana) two miles from the Observatory, which connects with the Chicago, Milwaukee & St Paul Railway (Chicago & Madison line) at Walworth, Wis, and with the Chicago & Northwestern Railway (main line to Minneapolis) at Harvard, Ill, twelve miles distant

feet above the ground. The length of the telescope is 62 feet, and spectroscopic attachments can be added which increase the length by nearly 10 feet. The instrument can be readily moved by hand, in spite of the weight, 6 tons, of the tube, and that (total 20 tons) of all moving parts. Electric motors are available, however, by which the instrument can be quickly pointed toward any part of the sky. For spectroscopic and photographic work it is necessary to correct the position of the telescope by very slight motions, for which additional motors, controlled by the observer's hand, are provided. With these the eye-end can be moved by an amount as small as $\frac{1}{100}$ inch, and stopped at the position desired. The driving-clock, by which the telescope is made to follow the stars, occupies a small room in the top of the pier. Its performance has been remarkably satisfactory during the decade it has now been in use, at temperatures ranging from $-25°$ F. to $+90°$ F., and under great variations as to humidity. When the driving-weight reaches the end of its descent within the pier, it touches a button, turning on a switch, and the clock is wound up by a motor. By shifting a gear, the rate of the clock may be changed to make the telescope follow either the sun, moon, or stars. Finely graduated circles were provided for reading off the position of the telescope, but in practice the two coarse circles, five and eight feet in diameter, with large graduations visible from the floor, have been found sufficient and quickest in operation. For sighting the instrument a finder of 4 inches aperture is provided, and for the spectrograph a small finder of 60 feet focus, without tube, is used.

The optical quality of the telescope is of the highest order; under favorable atmospheric conditions very faint objects have been seen with it, and close stars separated. A German authority who has investigated the quality of many objectives, large and small, by means of photographs taken inside and outside of the focus, places this at the head of the list of those thus far studied. A great telescope, of long focal length, makes very exacting demands upon the atmosphere, far more than is the case with a smaller instrument. Consequently a casual visitor would very probably be disappointed not to find a higher magnification employed on an average night. In a more favorable climate, such as that of California, the performance would undoubtedly be much better than here.[1] In the work done with the spectrograph the steadiness of the air is not of as much consequence, as unfavorable conditions merely protract the exposure without injuring the quality of the plate. For visual observations of objects which are sufficiently bright, a large iris diaphragm has recently been constructed in the shop, by which any desired part of the whole aperture may be used.

The operation of the great telescope is greatly facilitated by the admirable arrangements of the dome and rising floor, designed and constructed by Warner & Swasey. In order to have ample room for attaching spectroscopes and other apparatus, it was necessary that the dome should be large, and its diameter is

---

[1] *Weather at Lake Geneva.*—A record is kept of the number of hours during which the 40-inch telescope is in use at night It should be understood that except for cloudy skies it is employed throughout every night of the year. In 1908, a very clear year, it was in use during over 1880 hours, or about 52 per cent. of the total night hours. It is doubtful if many European observatories could give as good a record of clearness.

90 feet. - It is turned on 36 wheels by an electric motor which actuates an endless wire rope extending around the whole dome. The controller is on the rising floor just at the north side of the pier. Six minutes are required for turning the dome completely around. The rising floor is 75 feet in diameter, and is supported by four cables running over four sheaves just under the upper balcony and connecting with heavy counterpoises which balance a large part of the weight of the floor (37½ tons). A powerful motor, with its controller conveniently near to that for turning the dome, supplies the power. The floor moves through a range of 23 feet from the lower to the upper balcony. It has to be nearly at its lowest point when an observer is looking at a star near the zenith. In intermediate positions the floor can be instantly adapted to the height of the eyepiece of the telescope. The opening in the dome through which the sky is observed is 13 feet wide, and it is closed by shutters 85 feet long. These are so well constructed that they can be very easily moved by hand. Canvas screens, which may be raised over part of the opening, serve to break the force of the wind.

ACCESSORIES AND LINES OF RESEARCH

The principal accessories of the telescope and the lines of work done with each may now be mentioned. The filar micrometer is attached at the eye-end for nearly all visual observations. It was presented by Mr. Yerkes, and constructed by Warner & Swasey. It enables the observer to determine very exactly the angular separation of two objects (or points) both of which are visible in the eyepiece at the same time, and to fix the angle made by the line joining the points with the north-and-south direction. The micrometer has during the past ten years been in use for a little more than one-half the time, generally three or four nights per week. With it Mr. Burnham has made a great number of observations, most of which were utilized in his monumental catalogue of double stars published in 1906. Mr. Barnard has similarly measured with great skill and persistence the positions of many stars in a number of the star clusters, of the fainter satellites of our system, of nebulae and faint comets, and has observed the most interesting objects suitable for visual examination.

A very important method of using the telescope as a camera was successfully applied by Mr. G. W. Ritchey, principally in 1900 and 1901. The photographs obtained by him of the moon, star clusters, and some nebulae were at once recognized as among the very best thus far secured. For this service there is substituted, in place of the eyepiece, a plate-holder, which can be very delicately moved either north and south or east and west, and can thus follow the slight oscillations of the object due to the unsteadiness of our atmosphere during the exposure. The crown and flint lenses of the 40-inch objective were of course so figured as to be achromatic for the rays to which the eye is most sensitive, the yellow and green. By the interposition of a color-filter cutting off the blue and violet rays, which are not simultaneously in focus with the visual rays, and by the use of isochromatic plates, which are sensitized for the visual rays, a sharp photograph is obtained.

Fig. 3.—Lunar Crater Theophilus and Surrounding Region

Photographed with the 40-inch refractor in less than one-half second (Ritchey). The diameter of the large crater is 64 miles; its walls rise to a height of 18,000 feet, and the central peaks (notice their shadows cast toward the left) are about 6,000 feet high.

A negative of a star cluster, or other object, secured in this way when atmospheric conditions are particularly good, is then measured under a microscope in the laboratory, and even weeks of work can be profitably spent upon a single fine negative.

During 1904 and 1905 a large number of stellar photographs were thus obtained by Mr. Schlesinger, who measured them most carefully in order to determine from the plates the distances from the sun of certain selected stars. The results are not yet published, but they are probably the most accurate thus far attained, and much superior to any which could be secured visually.

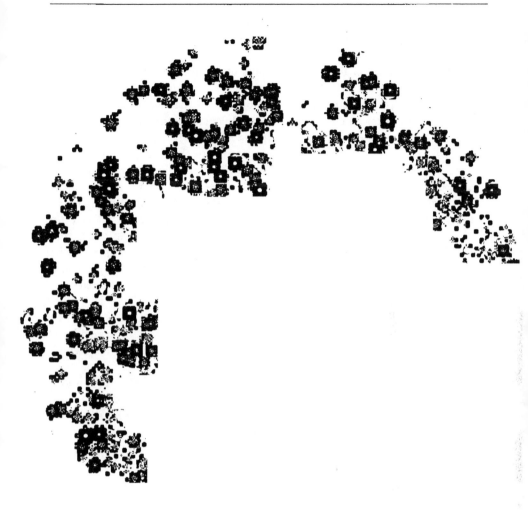

FIG. 4.—STAR CLUSTER IN HERCULES (Messier 13)

Photographed with the 40-inch refractor by Ritchey with an exposure of three hours. Over three thousand stars are visible on the original negative. The above field of view, including this group of suns, has an angular diameter in the sky one-third that of the moon. They are enormously distant from the earth and widely separated from each other: probably no two which can be distinguished separately are nearer each other than one million million miles.

Another attachment of the telescope is a stellar photometer with which the brightness of the stars is measured. Observations of many faint stars have thus been made by Mr. Parkhurst.

About one-third of the nocturnal hours of the telescope, generally two nights per week, are devoted to stellar spectroscopy. The instrument, which was largely constructed in the shops of the Observatory, is designated as the Bruce spectrograph, the funds for its construction and early operation having been contributed by the late Miss Catherine W. Bruce, of New York. A photograph of the spectrum of a star, accompanied by a comparison spectrum on the same plate from metallic terminals vaporized by the passage of sparks from an induction coil in the dome, can be measured under a microscope, and the speed of the star in the line of sight (averaging about 10 miles per second) can be inferred from the displacements of the star's lines from the position of the lines due to the spark. About 4,000 photographs, or spectrograms, have been obtained with the instrument, chiefly by Messrs. Frost, Ellerman, Adams, Barrett, and Lee. Interesting results have come from the measurement of the spectrograms, and some 70 spectroscopic binaries, or double stars so close to each other that they can be separated only in this spectroscopic manner, have been detected.

An especially important use of the great refractor has been in the study of the sun. The attachment generally used is the spectroheliograph, so named by Mr. Hale, its inventor, when he brought it into successful operation at the Kenwood Observatory.

The instrument regularly in use here for the past six years is known as the Rumford spectroheliograph. It is a heavy spectrograph (700 lbs.) with which, by the motion of the plate while the sun's image is passing across the slit, pictures of the sun's surface and surrounding prominences may be secured in monochromatic light, from some one line of calcium, or hydrogen, or other element, which may be particularly active on the sun. Adjustments can be so made that the photograph will depict either the lower, intermediate, or upper levels of the incandescent vapors that lie above the surface. The scarlet radiation of hydrogen shows in a marked degree the vortices forming around a sun-spot (which may be regarded as a storm in the sun's atmosphere), as was first found by Mr. Hale with his fine apparatus installed on Mt. Wilson, California. Visual and photographic studies of the sun are made with other spectroscopes, and direct photographs are also taken with the great telescope. This work was at first carried on by Messrs. Hale and Ellerman, and after their departure in 1904 by Mr. Fox.

OTHER TELESCOPES AND APPARATUS

The southeast dome, 30 feet in diameter, contains the two-foot reflector, an instrument quite as powerful for some purposes as the 40-inch telescope itself. The silvered concave mirror was made in the optical shop by Mr. Ritchey, who also supervised the construction of the mounting in the Observatory machine shop. The figure of the mirror is very perfect, and remarkably fine photographs

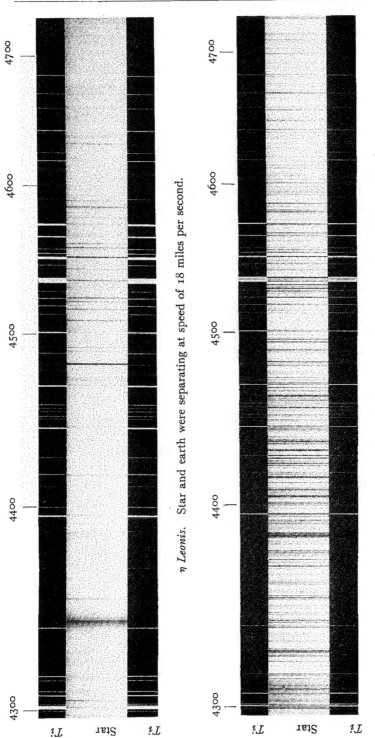

η *Leonis.*   Star and earth were separating at speed of 18 miles per second.

α *Boötis (Arcturus).*   Star and earth were approaching at speed of 11 miles per second.

FIG. 5.—STELLAR SPECTRA PHOTOGRAPHED WITH THE BRUCE SPECTROGRAPH (Frost and Adams)

The white lines are due to titanium vaporized by an electric spark in front of the spectrograph.   They furnish reference points on every plate.   The dark lines on a light background are in the spectrum of the star.   From the lack of perfect coincidence of the titanium lines in spark and star, the velocity of the star may be inferred, after positions of lines have been measured to one twenty-five-thousandth of an inch.   After correcting for the earth's velocity (in its orbit around the sun), the above stars are found to have speeds of 2 miles per sec. (recession) and 3 miles per sec. (approach), respectively.

of nebulae have been obtained with it, principally by Mr. Ritchey. It has the great advantage over a refractor that the rays of all colors are united in the same focus, here 8 feet from the surface of the mirror. A star of the ninth magnitude (having less than one-thousandth of the brightness of a first-magnitude star) can be photographed with it in one second, and the instrument is exclusively used

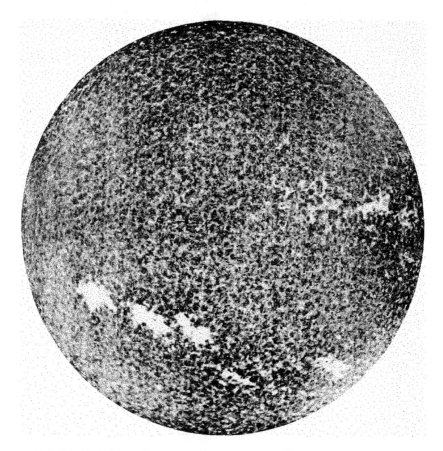

FIG. 6.—THE SUN, PHOTOGRAPHED WITH THE SPECTROHELIOGRAPH ON
AUGUST 12, 1903 (Hale and Ellerman)

The white blotches are areas of intensely brilliant calcium vapor, which would be invisible on an ordinary photograph.

photographically. With an exposure of three hours, vast numbers of stars will impress their images on the plate, many of which cannot be seen at all with the 40-inch refractor. Some work on stellar spectra has also been done with the instrument.

The northeast dome contains the Kenwood equatorial refractor of 12 inches aperture, which, together with much valuable equipment, was given to the Uni-

versity by Mr. William E. Hale, and moved here from its former location at
Mr. Hale's residence on Drexel Boulevard, Chicago. The instrument has two
objectives, one for visual work, the other for photography, both made by Mr.
J. A. Brashear. The mounting is by Warner & Swasey. An excellent filar
micrometer, by the last-named firm, was purchased for this telescope in 1906.
The old stellar spectrograph, provided by Mr. Yerkes in the original equipment
for the 40-inch telescope, has recently been adapted for use in solar spectroscopy

FIG. 7.—COMPOSITE OF TWO PLATES MADE ON AUGUST 14, 1907 (FOX)
Showing calcium vapor as prominences at sun's edge and on the disk.

with the 12-inch telescope. The telescope has been used for observations of
comets, double stars, variable stars, for direct photographs of the sun, stellar
photometry, etc.

On the main roof, between the two small domes, is housed the Brashear
comet-seeker, of six inches aperture. In the meridian room at the east end of the
building is located the broken-tube transit of 3 inches aperture, made in the
machine shop. It is used for determining the time.

At the north window of the short corridor leading to the meridian room are
the barometer, maximum and minimum thermometers, and the thermograph.
Observations are made twice a day, solely for the purposes of the Observatory,
and not for publication. A thermograph has also lately been installed on the

FIG. 8.—THE 24-INCH REFLECTOR, OCCUPYING THE S. E. DOME

The silvered glass mirror was made by Mr. Ritchey in the optical shop, the mounting in the machine shop of the Observatory. It will photograph stars which cannot be seen with the great refractor.

north side of the great dome on the outside balcony, to give a record of temperature for correction for the atmospheric refraction. No seismographs for recording earthquakes, nor instruments for determining the state of the earth's magnetism are maintained at the Observatory.

In the basement has been set up an apparatus devised by Professor Hale, by which the photographs of the sun taken with the spectroheliograph can be optically projected upon a steel globe ruled with parallels and meridians, so that the positions of significant markings on the sun can be easily read off without trigonometrical computation.

An important adjunct of an astrophysical observatory is its spectroscopic laboratory, where experiments may be made to match in some degree the conditions found in celestial objects, in running to earth unknown elements in their spectra, and in various researches calculated to assist in the interpretation of the phenomena of the heavenly bodies. Unfortunately it has not been possible to obtain funds for keeping a physicist steadily employed. During two or three years, principally from a gift of about $4,000 made by Mr. Yerkes, such work was in progress in the well-situated and fairly well-equipped laboratory in the basement. Extensive studies were made by Mr. N. A. Kent, in collaboration with Mr. Hale, on the effect of heavy pressure upon the spectrum of the spark; by Mr. Kent on the comparison of wave-length in the spectrum of the same element when rendered luminous in spark and in the arc; and by Mr. H. M. Reese on the enhanced lines in spark spectra.

In view of the very general use of photography in modern astronomy, a laboratory for photographic research is important for an institution like this. In the years from 1903 to the end of 1908 such a laboratory was in successful operation under the charge of Mr. R. J. Wallace. Particularly during the last three years, under a grant to the Observatory from Mr. George Eastman, of Rochester, N. Y., Mr. Wallace was able to devote his time to research on various problems in photography, including the adaptation and sensitization of plates for particular purposes. Aside from the improvement of the photographic routine in the different departments of the Observatory, these investigations resulted in important papers published in the *Astrophysical Journal* and elsewhere.

The great optical works at Jena, having the firm-name Carl Zeiss, have rendered great service in the development of microscopes and apparatus for measuring photographs. Several are owned by the Observatory, the most important being known as the "Stereo-comparator" and "Blink-mikroskop," purchased in 1907. This is a massive instrument in which two different negatives can be examined, either at the same time or in instantaneous succession, so that any changes in position, size, or intensity of the images is at once noticeable to the eye, and discoveries are greatly facilitated. Accurate measurements of various kinds can also be made with this apparatus. Another valuable accessory is the Hartmann spectrocomparator, acquired at the same time, with which two negatives of spectra can be simultaneously compared and measured against each other. Several other measuring machines, some of them made in the Observatory shop, are in use for the precise study of plates. In the computing room, calculating machines made in Europe, capable of handling 18 figures, are principally employed for multiplication and division.

FIG. 9.—THE GREAT NEBULA IN ORION, PHOTOGRAPHED WITH THE 24-INCH REFLECTOR (Ritchey)

This inconceivably large mass of gas, chiefly hydrogen, helium, and nebulum (not yet found in the earth), is a feature of the constellation Orion.

As the original equipment of the Observatory was chiefly limited to the 40-inch and 12-inch telescopes, much remained to be done in fitting out the institution for its best efficiency. The trustees of the University appreciated the wisdom of the recommendation of the director, and provided from the start for maintaining an instrument-maker and machine shop, the machines of which were chiefly donated from the Kenwood Observatory. In this way the equipment has been gradually increased, kept in repair, and perfected. At times, when special gifts were available, three men have been employed in the shop. The principal appa-

Fig. 10.—Nebula in Triangulum (Messier 33), Photographed with the 24-inch Reflector (Ritchey)

The spiral structure can be readily seen. Its distance from the earth is not less than that of the surrounding stars, which, it should be recalled, are all suns, presumably not smaller and perhaps much larger than our own sun.

ratus constructed have been these: the two-foot reflector, the Snow horizontal telescope, the Bruce spectrograph, the transit instrument, a spectroheliograph, a coelostat for eclipse use, a quartz spectrograph for the two-foot reflector, and many smaller instruments and attachments. In the carpenter shop a great number of patterns have been made, and special cabinets of various kinds have been constructed for negatives, charts, and other valuables.

The optical shop was a particular feature of the early operation of the Observatory. This was in charge of Mr. Ritchey, and by himself, or under his direction, a great number of mirrors of the highest optical quality, flat and concave, including the two-foot reflector, were made. During the first two years the work was chiefly on the great five-foot mirror, under a gift from Mr. William E. Hale. As the University was finally unable to accept the terms of gift and to provide a mounting and building for this instrument, it later became a part of the equipment of the Mt. Wilson Solar Observatory of the Carnegie Institution at Pasadena. After being finished at the Pasadena optical shop, and being provided with a fine mounting and dome on Mt. Wilson, one of the best sites for astronomical work yet found, it is now, at the beginning of 1909, just coming into operation—undoubtedly the most perfect instrument of its kind in the world.

HORIZONTAL TELESCOPE

Owing to the great convenience of operation of a horizontal coelostat telescope for solar work where massive spectroheliographs must be attached and removed, Mr. Hale early began the construction of such an instrument. The first one had barely been completed, in a temporary wooden and paper structure outside of the Observatory, when it was burned (December 22, 1902) from the sparks of the high-potential apparatus used for the comparison spectrum. A 30-inch plane mirror and a 24-inch concave mirror, made in the optical shop, were destroyed, together with the driving-clock and all but the heavy castings of the coelostat. A few months later Miss Helen M. Snow, of Chicago, made a gift of $10,000 in honor of her father, the late George W. Snow, for reconstructing the instrument. A wooden building was erected 600 feet north of the Observatory, and by great activity in the optical and machine shops it was possible formally to open the Snow telescope in October, 1903.

Having been interested in the possibilities of solar research under the more perfect atmospheric conditions of southern California, Mr. Hale, in the next year, obtained a grant of $10,000 from the Carnegie Institution of Washington for an expedition for solar research from the Yerkes Observatory to Mt. Wilson, Cal. The Snow telescope was safely transported thither and mounted in a steel and canvas building, which experience showed to be more suitable than the wooden structure. The results obtained were so satisfactory and so promising for the future, that the Carnegie Institution decided to establish its own observatory there. An arrangement satisfactory to the donor and all others interested was later made whereby the Snow telescope was purchased by the Solar Observatory, the money thus returned to the University of Chicago being used to buy the stereo-comparator, and $5,000 being established as the Snow fund, from the interest of which instrumental equipment can be purchased. This is the sole endowment fund thus far provided specifically for the use of the Observatory. The greatest loss to the Observatory was the departure (in 1904) to the new field of activity of Mr. Hale, Mr. Ellerman, Mr. Ritchey, and Mr. Adams.

In 1908 the eclipse coelostat was mounted in the Snow building in connection with an auto-collimating spectrograph, and a powerful instrument for research in the solar spectrum was secured at small additional cost.

BRUCE PHOTOGRAPHIC TELESCOPE

In 1897 in response to a presentation of the advantages of such an instrument by Mr. Barnard, Miss Catherine W. Bruce, of New York, placed in his hands, as a gift to the University of Chicago, the sum of $7,000 for the construction of a telescope especially designed for the photography of large areas and extended objects in the sky, particularly the Milky Way and comets—a field in which Mr . Barnard has been a particularly successful pioneer. A number of years were spent in the endeavor to secure a lens which should fully meet the ideal requirements. A fine 10-inch doublet, constructed by Brashear, was finally accepted, in 1900, and a special mounting was designed and built by Warner & Swasey. The instrument was set up in April, 1904, in a neat wooden building (erected from interest of the Bruce fund), 350 feet southwest of the great dome of the Observatory. In addition to the 10-inch (focal length 50 inches), the mounting also carries a 6¼-inch Voigtländer doublet, a 5-inch visual guiding telescope, and generally a 3½ inch portrait lens. Many photographs of the Milky Way, showing good definition over a field of 50 square degrees, have been obtained by Mr. Barnard. In 1905 he spent nine months at Mt. Wilson (altitude nearly 6,000 ft.) in photographing the Milky Way, particularly those southern portions not to be reached at Lake Geneva. This trip was made possible by a grant from Mr. J. D. Hooker, of Los Angeles. The Carnegie Institution of Washington has lately made a grant for the purpose of publishing an atlas or photogravure reproductions of these pictures of the Milky Way, and the engravers are now engaged in attempts to perfect their work. Of the comet Morehouse, discovered on plates taken with the Bruce telescope by Mr. Morehouse on September 1, 1908, Mr. Barnard secured no less than 350 negatives.

Another photographic lens of special value is the Zeiss doublet, made of glass particularly transparent to ultra-violet rays, of 5.7 inches aperture and 32 inches focal length. The stellar images it produced when slightly out of focus were found to be disks of remarkably uniform intensity. By measuring the intensity of these disks under the Hartman "surface photometer" Mr. Parkhurst has attained an exceedingly high degree of accuracy in stellar photometry. An objective-prism of the same transparent glass and of the same aperture as the lens is provided for the doublet and makes possible the study of the spectra of the stars which are within the field of view. The Zeiss doublet is used on a mounting and in a small building belonging to Mr. Parkhurst, located 1,000 feet north of the Observatory.

Fig. 11.—The Bruce Photographic Telescope Temporarily Set Up in the Corridor of the Yerkes Observatory

It is now located in the small dome S. W. from the Observatory. With this Professor Barnard has obtained many unsurpassed photographs of the Milky Way and of comets.

FIG. 12.—A SMALL PORTION OF THE MILKY WAY IN CEPHEUS PHOTOGRAPHED
WITH THE BRUCE TELESCOPE ON SEPTEMBER 11, 1904 (Barnard)
EXPOSURE 6 HOURS AND 40 MINUTES

As will be seen in this picture, the Milky Way consists of innumerable faint stars
(in reality distant suns), separately indistinguishable with the naked eye, which are
distributed irregularly in the sky. Only three or four of the stars on this plate can
be seen with the naked eye.

The power-house, which also furnishes the steam heat (largely exhaust) for the main building, is situated about 750 feet northeast from the center of the large dome. It was included in the original gift of Mr. Yerkes. Its equipment is in duplicate, consisting of 8×10 Ideal engines, directly connected with Siemens & Halske dynamos, of capacity 200 amperes at 125 volts, which furnish current for the motors in the shops, for the great dome and rising floor, for the 30-foot dome, and for all the general lighting and dark-room lamps of the Observatory. The water for all purposes of the institution is obtained from the lake by an electric pump at the shore operated by current from the power-house.

Questions are so often raised as to the cost of the equipment that a brief state-ment may be given here. The land originally given was valued at $30,000. The cost of the completed object-glass of the great refractor was $66,000; of the telescope mounting itself, $55,000; of the dome and rising floor, $45,000; and of the remainder of the Observatory building, including the southeast dome and the power-house and its equipment, about $150,000. The instruments and equipment of the Kenwood Observatory, with its dome, the gift of Mr. William E. Hale and his son, were valued at $30,000. The Bruce telescope and building cost about $9,000; the Snow telescope and building, $10,000.

In response to the numerous calls for lantern slides, transparencies, and prints from astronomical negatives made at this Observatory, it became necessary several years ago to arrange for supplying them to teachers, lecturers, and others interested. A young man is employed in preparing these reproductions, which are sold at what is actually less than cost, through the medium of the University of Chicago Press. A catalogue of such slides, etc., and two appendices, includ-ing a total of about 500 subjects, have been printed, and are sent on request by the University of Chicago Press.

PUBLICATIONS

Current results of the investigations at the Observatory are published in various journals, American and foreign. Those of an astrophysical character generally appear in the *Astrophysical Journal*, of which the director of the Observ-atory is managing editor, published ten times a year by the University of Chicago Press. The Carnegie Institution of Washington has published several separate volumes by members of the staff, two of which have been already referred to. Others are: *Researches in Stellar Photometry* by J. A. Parkhurst; *The Rotation Period of the Sun* by George E. Hale and Philip Fox. The University authorities have not yet been able to provide a regular fund for the issue of the formal quarto *Publications of the Yerkes Observatory*, and these are therefore much in arrears. Only two volumes have so far been issued: I (1900), pp. 296, *A General Catalogue of 1290 Double Stars Discovered from 1871 to 1899* by S. W. Burnham; II (1904), pp. 413, papers by Messrs. Burnham; Barnard; F. R. Moulton; Frost and Adams; Hale, Ellerman, and Parkhurst; Ritchey; and Laves.

A valuable work by Professor Hale entitled *The Study of Stellar Evolution* has recently been published by the University of Chicago Press. It describes in

FIG. 13.—COMET MOREHOUSE AS PHOTOGRAPHED WITH BRUCE TELESCOPE ON OCTOBER 14, 1908 (Barnard)

The plate was exposed for seventy minutes. The observer kept, the telescope constantly pointed on the comet, following its motion among the stars which caused the elongation of the star images. The tail of the comet, as shown on this plate, was about 14 million miles long. It was composed of luminous vapors, chiefly of carbon and nitrogen.

particular the methods and results of research at the Yerkes and Mount Wilson observatories in their relation to the general investigation of the problem of stellar evolution.

INSTRUCTION

Undergraduate instruction in astronomy is not given at the Observatory. This is provided at the University, together with thorough courses in theoretical astronomy and celestial mechanics taught by two professors and one instructor. Graduate students competent in observational work in astronomy and astrophysics are welcome and may become Fellows under the usual University regulations.

All candidates for the doctor's degree in the department are required to work at least one quarter at the Observatory. Astronomers from other institutions frequently spend the summer here as volunteer research assistants.

The Observatory library contains at present about 3,000 volumes, and 3,000 pamphlets. About 80 scientific magazines and journals are regularly received, principally in exchange for the *Astrophysical Journal.*

The pressure for time for scientific use has made it impossible to permit visitors to see through the telescopes. Opportunity is given, however, for them to inspect the Observatory and the great refractor, on Saturdays: from June 1 to September 30 between 1:30 and 4:30 P. M.; during the remainder of the year, between 10 and 12 A. M. A member of the staff demonstrates the operation of the large telescope and explains the work of the Observatory. Several thousand visitors annually avail themselves of this opportunity.

The staff of the Observatory is constituted as follows (April 1, 1909):

EDWIN B. FROST, Professor of Astrophysics and Director.

SHERBURNE W. BURNHAM, Professor of Practical Astronomy.

EDWARD E. BARNARD, Professor of Practical Astronomy.

JOHN A. PARKHURST, Instructor in Practical Astronomy.

STORRS B. BARRETT, Secretary and Librarian.

PHILIP FOX, Instructor in Astrophysics.

OLIVER J. LEE, Computer.

MARY R. CALVERT, Computer.

MARY F. WENTWORTH, Stenographer.

FRANK R. SULLIVAN, Engineer in charge of 40-inch telescope

OSCAR E. ROMARE, Instrument Maker.

HENRY J. FOOTE, Carpenter.

WILFRED BEGUELIN, Lantern Slides.

DIEDRICH J. OETJEN, Day Engineer.

LOUIS F. CLAY, Night Engineer.